CARING FOR AFRICAN SPURRED TORTOISE

A BEGINNER'S GUIDE TO KEEPING THIS UNIQUE PET

DR MORRIS HART

Table of Contents

Introduction

Scientifically called as Geochelone sulcata, African spurred tortoises are fascinating animals that are gaining popularity as pets. These tortoises, which are native to the dry parts of sub-Saharan Africa, have unique characteristics and habits that make them fascinating pets for any home. We explore the natural environment, physical traits, life cycle, and reasons why African spurred tortoises are great pets for devoted aficionados in this extensive guide to the species.

- Organic Environment

Native to sub-Saharan Africa, African spurred tortoises are arid, scorching savannas and grasslands. They are mostly found in Senegal, Mauritania, Mali, Niger, Chad, Sudan, Ethiopia, and Eritrea, among other nations. They live in a range of habitats in these areas, from semideserts to scrublands, where they have evolved to withstand aridity and a lack of vegetation.

- Physical attributes

The remarkable size of African spurred tortoises is one of their most remarkable characteristics. With adult lengths of 24 to 30 inches and weights of 70 to 100 pounds or more, they are among the largest species of tortoises in the world. Their domed, usually light brown to yellowish-colored shells are ornamented with characteristic elevated scutes and growth rings. By retaining heat throughout the cooler months, these scutes shield the tortoise from predators and aid in controlling body temperature.

Their strong, robust limbs are designed to allow them to move easily across their natural habitat. Strong, robust legs and columnar feet with sharp claws for gripping and digging on uneven surfaces are characteristics of African spurred tortoises. Their heads are small in relation to their bodies, with a blunt nose and strong teeth adapted for feeding on hardy flora, and their tails are short and stubby.

- Lifespan

When given the right care, African spurred tortoises can live for up to 50 years in captivity. They mature sexually between the ages of 15 and 20, depending on a variety of factors including genetics, food, and habitat. Males engage in courtship displays to entice females during mating, which usually takes place during the rainy season. Following mating, females create small nests in the ground where they lay clutches of eggs. Typically, these nests are dug during the cooler months to mark the beginning of the dry season. After several months of incubation, hatchlings are prepared to leave their eggs and begin their journey of growth and development.

- The Reasons African Spurred Tortoises Are Wonderful Pets

African spurred tortoises are becoming more and more popular as pets because of their placid disposition, minimal upkeep needs, and distinct personalities. They make good companions for families with young children since, in contrast to certain other reptiles, they are usually peaceful and tolerant of handling. Their slow metabolism means they don't need to be fed frequently, and their herbivorous diet of grasses, leafy greens, and occasionally fruits and vegetables is reasonably easy to give. Furthermore, because of their lengthy lives, they can be treasured family members for many years to come, offering their owners company and pleasure.

Spurred tortoises from Africa are amazing animals that provide a window into the beauties of the natural world. With their distinct beauty and appeal, these magnificent reptiles have won over many hearts since they first appeared in the African savannas and have since become popular pets in homes all over the world. A gratifying path of caring for and appreciating these amazing animals for years to come can be taken by lovers who understand their natural environment, physical traits, lifecycle, and why they make such great pets.

Chapter 1

Housing Needs: Designing the Perfect Environment

For African spurred tortoises to be healthy and happy, an appropriate habitat must be provided. Native to sub-Saharan Africa, these tortoises are semi-arid dwellers with certain environmental requirements that must be satisfied in captivity in order to guarantee their long-term survival. We cover everything from enclosure size and substrate to lighting and temperature and humidity levels to enrichment to support the physical and mental well-being of African spurred tortoises in this extensive guide on housing requirements.

- Size and Design of Enclosures

Bigger is usually preferable when it comes to housing African spurred tortoises. These tortoises need lots of room to move about, feed, and exercise in order to replicate as much of their original environment as possible. Although an expansive outdoor cage is preferable, offering the greatest enclosure feasible is crucial if inside living is required due to climate or space restrictions.

It is advised that indoor enclosures be at least 8 feet by 4 feet for a single adult tortoise, with more room required for each additional tortoise. Aim for at least 100 square feet per tortoise in outdoor enclosures, with sturdy fencing to keep them safe from predators and to prevent escape.

To discourage climbing and escape attempts, the enclosure should be built with walls that are at least 18 inches high, made of durable materials like concrete,

PVC, or wood. To preserve the quality of the air and avoid respiratory problems, make sure the enclosure has adequate ventilation.

- Substance

For African spurred tortoises, selecting the appropriate substrate is essential since it offers a pleasant surface for walking and burrowing and aids in maintaining the ideal humidity levels. For traction and moisture retention, a base composed of sand, coconut coir, and organic topsoil is effective.

Steer clear of extremely rough or coarse substrates as they may injure the tortoise's delicate underside. Additionally, as they might be poisonous to reptiles, surfaces like pine bark or cedar shavings should be avoided.

Make sure the substrate is at least 6 inches deep to facilitate digging and nesting activities. To keep the substrate clean and stop bacterial growth, replace it completely every few months and spot-clean it frequently to remove waste and uneaten food.

- Humidity and Temperature Ranges

African spurred tortoises need proper temperature and humidity levels to be maintained for their health and welfare. To efficiently regulate their body temperature, these reptiles need a warm, dry environment with access to basking places and cooler retreats.

During the day, create a basking area beneath a heat lamp or ceramic heat emitter that is between 90 and 95°F (32 and 35°C). During the day, the enclosure's ambient temperature should be maintained between 75 and 85°F (24 and 29°C), and at night, it shouldn't drop below 70°F (21°C).

Moderate humidity is defined as having a relative humidity between 40 and 60%. Use a spray bottle to mist the enclosure every day to maintain humidity, or install a humidifier if needed. For the tortoises' comfort, provide a humidity hide with moist sphagnum moss and a shallow water dish for soaking.

- Luminance

For African spurred tortoises to remain healthy and happy, adequate illumination is necessary. For indoor enclosures, additional UVB illumination is required in addition to natural sunlight, which delivers vital UVB radiation for the production of vitamin D and the metabolism of calcium.

Cover the enclosure's basking area with a UVB fluorescent light bulb (such as a ReptiSun 5.0 or 10.0) with a wavelength of 290 to 310 nm. To guarantee proper exposure, place the lightbulb between 12 and 18 inches above the tortoise's shell. To keep the UVB output at its best, replace the bulb every six to twelve months.

Provide a heat lamp or ceramic heat emitter in addition to UVB lighting to generate a basking area that is between 90 and 95°F (32 and 35°C). Because of this, tortoises are able to thermoregulate by alternating between the enclosure's warmer and colder sections as needed.

- Enhancement

For African spurred tortoises, enrichment is a vital component of housing because it offers opportunity for natural behaviors including digging, foraging, and exploring as well as cerebral stimulation. To keep tortoises occupied and active, include a range of environmental enrichment objects in the enclosure.

Provide hiding places where tortoises can hide and feel safe, like hollowed-out boulders, plant pots, or half logs. Add barriers to promote exploration and exercise, as well as climbing and basking spots made of branches, rocks, and other natural objects.

Provide a range of plants, such as grasses, leafy greens, and edible weeds, for tortoises to graze on. Directly planting edible plants into the substrate can promote natural foraging behaviors and offer further enrichment.

The health, welfare, and general quality of life of African spurred tortoises depend on the provision of their perfect habitat. Owners of tortoises can establish a safe and stimulating habitat for their pets to thrive in by making sure their enclosure satisfies their individual needs regarding size, substrate, lighting, temperature and humidity levels, and enrichment. African spurred tortoises make wonderful family pets that can live long, healthy lives if given the right care and attention to their housing needs.

Chapter 2

Tips for Eating and Nutritious Food for Your African Spurred Tortoise

Maintaining the general health and wellbeing of your African spurred tortoise requires feeding it a nutritious food that is well-balanced. African spurred tortoises are herbivores with particular dietary needs, and they depend on a range of plant-based meals to meet their dietary requirements. We examine the food preferences of African spurred tortoises, foods that are suggested, feeding frequency, supplementation, and typical feeding errors to avoid in this extensive guide.

- Food Preferences

African spurred tortoises are mostly herbivores, which means that the majority of their diet consists of plants.

They feed on a range of grasses, leafy greens, flowers, and succulent plants that grow in their natural habitat when they are in the wild. In order to effectively digest fibrous vegetation and extract nutrients from hard, fibrous plant material, they have evolved specialized digestive systems.

In the wild, they might sometimes eat tiny amounts of fruit, such ripe fruits or fallen berries, but because fruit contains a lot of sugar, it should only be given in small amounts to captive animals. Furthermore, because they are strict herbivores, African spurred tortoises shouldn't be fed animal-based proteins like meat or insects because these could cause digestive problems and other health problems.

- Suggested Foods

To ensure that African spurred tortoises receive the nutrition they require, a varied and well-balanced diet is necessary. To make sure they get a mix of vitamins, minerals, and fiber, offer a variety of grasses, weeds, and leafy greens. Among the foods that are advised are:

Grasses: Wheatgrass, Bermuda grass, orchard grass, and Timothy hay are all great sources of dietary fiber that aid in a healthy digestive system.

Leafy Greens: Rich in vitamins A and K, calcium, and other necessary elements, dark greens like collard, dandelion, mustard, kale, and turnip greens are a great source of nutrition.

Edible Weeds: Native plants including sow thistle, plantain, chickweed, and clover offer a range of nutrients and natural foraging opportunities.

Vegetables: As occasional treats, serve a variety of vegetables in moderation, such as bell peppers, cucumbers, zucchini, squash, and carrots.

Flowers: Edible flowers can be given occasionally as treats. Examples of edible flowers are hibiscus, dandelion, nasturtium, and rose petals. These flowers bring enrichment.

- Frequency of Feeding

Compared to several other reptiles, African spurred tortoises have slow metabolisms and do not need to be fed frequently. Provide a variety of grasses, leafy greens, and vegetables to adult tortoises once a day to promote a balanced diet. More frequent feedings—up to twice a day—may help younger tortoises maintain their rapid growth and development.

Keep an eye on your tortoise's appetite and alter the frequency of feedings as necessary. Giving too much food can result in obesity and other health problems, so it's important to feed them in balance and refrain from giving them too many high-calorie goodies.

- Addendum

African spurred tortoises can thrive on a diversified diet consisting primarily of fresh, nutrient-rich foods, but supplementation may be required to make sure they get all the vital vitamins and minerals they require. A few times a week, dusting their meals with calcium powder and vitamin D3 will assist maintain healthy bone growth and development and help prevent metabolic bone disease.

Furthermore, giving a multivitamin supplement to reptiles once or twice a month will assist close any dietary gaps and promote general health. To determine the proper supplementation schedule for your tortoise, speak with a veterinarian who specializes in reptile care. Supplementation should never take the place of a well-balanced food, though.

- Typical Feeding Errors to Avoid

Although feeding African spurred tortoises might appear simple, there are a few frequent errors that owners should be aware of to protect the health and welfare of their animal companion:

Feeding an unbalanced diet: To make sure your tortoise gets all the vital nutrients it needs, try to feed a variety of plant-based foods rather than the same items over and over again.

Serving the wrong foods: Steer clear of high-sugar fruits, cereals, and processed foods as these can cause obesity and stomach problems. Consume only organic, plant-based foods that are similar to the diet that tortoises naturally would have in the wild.

Overfeeding: To avoid overfeeding and obesity, keep an eye on your tortoise's weight and modify the amount of food it receives as necessary. Food should be consumed in moderation if one wants to stay healthy and maintain a healthy weight.

Ignoring hydration: Make sure your tortoise always has access to clean, fresh water for bathing and drinking. Drinking enough water is crucial for healthy digestion and general wellbeing.

Maintaining the best possible health and wellbeing in your African spurred tortoise requires providing it a good, well-balanced diet. Tortoise owners may make sure their pets get all the vital nutrients they require for optimum health by giving them a range of grasses, leafy greens, veggies, and occasionally treats. They can also make sure their pets receive the right supplements and closely check the frequency of feedings. You may help guarantee a long and healthy life for your African spurred tortoise by avoiding frequent feeding errors and seeking advice from a veterinarian skilled in reptile care.

Chapter 3

Handling and Socialization: Interacting Safely with Your Pet

African spurred tortoises require handling and interaction in order for their owners to form strong bonds with them and give them opportunities for enrichment. These reptiles can nevertheless gain from gentle handling and positive relationships with their human caregivers, even though they might not be as engaged as some other pets. This thorough guide covers safe handling methods, socialization strategies for your tortoise, potential dangers to be aware of, and strategies for developing mutual respect and trust with your companion.

- Techniques for Safe Handling

It's crucial to put African spurred tortoises' safety and wellbeing first while reducing stress and discomfort when working with them. To guarantee that you and your tortoise have a great time together, use these safe handling techniques:

Approach gently and calmly to avoid startling your tortoise with abrupt movements or loud noises. Speaking quietly to reassure them, approach from the front or side so they can see you coming.

hold the shell: Always use both hands to hold your tortoise's shell when lifting it, one to support the carapace (top shell) and the other beneath the plastron (bottom shell). It can be harmful or stressful to pick up your tortoise by its limbs, tail, or head.

Lift softly and steadily: To prevent your tortoise from retreating into its shell, lift it slowly and steadily, without making any abrupt movements or jerks. To avoid tipping or imbalance, provide equal support for their weight.

Once your turtle has been raised, hold onto it firmly yet gently to keep it from slipping or falling. Squeezing too tightly can hurt or cause pain, so try to avoid doing so.

While occasional handling is helpful for bonding and socialization, avoid handling your tortoise excessively since this can lead to stress and interfere with its natural behavior patterns. Try to keep handling sessions brief and give your tortoise lots of downtime in between contact.

- Socialization Pointers

In order to assist your African spurred tortoise feel at ease and self-assured in a variety of settings, socialization entails introducing them to a variety of surroundings, stimuli, and experiences. To properly socialize your tortoise, adhere to these tips:

Start early: It's best to start socializing your tortoise when it's still a tiny hatchling or juvenile. Because they are more flexible and open to new experiences, young tortoises are simpler to socialize with.

Introduce gradually: To avoid overwhelm and lessen stress, acclimate your turtle to new situations, people, and animals gradually. As your tortoise gets more at ease and self-assured, gradually increase exposure by starting with brief, supervised interactions in a familiar and controlled setting.

Provide positive reinforcement: To promote desirable behaviors and foster trust with your tortoise, use positive reinforcement strategies such giving treats, praise, and favorite foods. During handling and socialization sessions, give rewards for calm and relaxed behavior to strengthen positive associations.

Repetition, consistency, and patience are all necessary for the progressive process of socialization. With patience, you may help your tortoise learn positive habits and gradually increase their exposure to new situations over time. Let them grow at their own speed.

Respect their boundaries: To ascertain your tortoise's comfort level and boundaries, observe their body language and clues. Give them space to escape and feel comfortable if they exhibit symptoms of stress or discomfort, such as hissing or retreating inside of themselves. Respect their limits.

- Possible Dangers to Steer Clear of

Although African spurred tortoises benefit from handling and socialization, it's important to be aware of potential risks and hazards to avoid:

Damage from improper handling: Mishandling your tortoise, such as dumping them or gripping them too firmly, can cause harm or discomfort to them. To reduce the chance of harm, always handle your tortoise gently and provide adequate support for its shell.

Contamination by pathogens: Your tortoise has a higher risk of illness if you handle it after handling other animals or if it comes into contact with polluted surfaces. To stop the spread of illness, carefully wash your hands both before and after handling your tortoise.

Stress-induced illness: Your tortoise may become unwell or experience a weakened immune system as a result of excessive handling or socialization, which can also cause stress and interfere with its natural behavioral patterns. Keep an eye out for indications of stress in your tortoise, such as appetite loss, fatigue, or behavioral abnormalities, and modify handling and socialization techniques as necessary.

Escapes and injuries: African spurred tortoises are robust, driven creatures that have the ability to break free from captivity or hurt themselves in strange places. To keep your tortoise safe and prevent escapes, always keep an eye on them during handling and socialization sessions.

- Establishing Mutual Respect and Trust

To promote your African spurred tortoise's general wellbeing and help you cultivate a pleasant relationship with them, you must first establish mutual respect and trust. To help you and your pet develop mutual respect and trust, consider the following advice:

Be dependable and consistent: Give your tortoise structure and predictability by following a regular feeding, handling, and care schedule. Provide a secure and supportive environment while being trustworthy in attending to their needs.

Recognize and honor their own preferences: Since each tortoise is different, its handling and socialization needs may vary as well. To gain your tortoise's trust and confidence, respect its own preferences and boundaries and modify your approach as necessary.

Spend quality time together: Set aside some time each day to engage in calm, constructive interactions with your tortoise, such as handling, feeding, or just watching them behave. Spending quality time together can help to develop your relationship because it takes time and effort to build a bond built on mutual respect and trust.

Provide opportunities for mental stimulation and enrichment: To keep your tortoise happy and occupied, provide it with mental stimulation opportunities and enrichment activities. To promote natural behaviors and stave off boredom, provide a range of environmental enrichment items, such as hiding places, climbing structures, and foraging opportunities.

Handling and socialization are crucial components of African spurred tortoise care, enabling owners to create lasting relationships and offer opportunities for enrichment. You can guarantee a happy and fulfilling experience for yourself and your pet by adhering to safe handling procedures, socialization advice, and awareness of potential hazards to avoid. It takes time and care to develop mutual respect and trust, but with commitment and perseverance, you may create a lifetime strong friendship with your African spurred tortoise.

Chapter 4

Health Care: Typical Problems and Proactive Steps

Proactive care and attention to the specific demands of African spurred tortoises are necessary to ensure their health and well-being. These reptiles are prone to a range of health problems that, if addressed, could negatively affect their quality of life, just like any other animal. This thorough guide covers typical health problems African spurred tortoises may face, proactive ways to maintain their health, warning indications of illness to look out for, and what to do if your tortoise needs medical assistance.

- Typical Health Concerns

When given the right care and management, African spurred tortoises are often hardy creatures. They might still be vulnerable to some health problems, some of which are more prevalent than others. You can spot difficulties early and take the necessary action if you are aware of these health disorders and their possible causes. Among the frequent health problems that African spurred tortoises face are:

Respiratory infections: Poor husbandry practices, such as inadequate ventilation, excessive humidity, or draft exposure, can lead to respiratory infections, a common health problem in African spurred tortoises. Lethargy, difficulty breathing, nasal discharge, and wheezing are possible symptoms.

Metabolic bone disease (MBD): Deficits in calcium and vitamin D3 can induce metabolic imbalances, weakening of the bones, and shell abnormalities. MBD is a dangerous disorder. Common contributing factors include an inappropriate diet, insufficient UVB lighting, and poor calcium supplements.

African spurred tortoises are susceptible to shell damage and rot due to falls, poor treatment, and predator attacks. These wounds may result in a fungal or bacterial infection of the shell called shell rot, which softens, discolors, and smells bad.

Infections with parasites: African spurred tortoises are susceptible to infections with internal and external parasites, including worms, mites, and ticks. These infections can cause poor eating, weight loss, lethargy, and skin irritation. To keep your turtle healthy, regular fecal inspections and preventative actions against parasites are crucial.

Eye infections: African spurred tortoises may develop conjunctivitis, or eye infections, as a result of inadequate hygiene, irritation from foreign objects or substrate, or underlying medical conditions. Eye discharge, red or swollen eyes, and trouble opening or closing the eyes are possible symptoms.

- Preventive Actions

African spurred tortoises need proactive care and attention to their food, environmental, and husbandry requirements in order to prevent health complications. You can maintain your tortoise's health and lower its chance of developing common health problems by taking preventative actions. Among the precautionary actions to think about are:

Giving your tortoise the right housing involves making sure it has a roomy, well-ventilated habitat with the right substrate, humidity, and temperature. Assign hiding places, places to bask, and clean water for soaking and drinking.

Providing a balanced diet: To make sure your tortoise gets all the vital nutrients they require, feed them a varied diet that includes grasses, leafy greens, vegetables, and occasionally treats. Steer clear of processed meals, high-sugar fruits, and grains as these can contribute to obesity and other health problems.

Supplementing with calcium and vitamin D3: You may help avoid metabolic bone disease and guarantee appropriate bone growth and development by dusting your tortoise's food several times a week with calcium powder that contains vitamin D3. To cover any dietary shortfalls, give a multivitamin supplement for reptiles once or twice a month.

Keeping your tortoise clean and clear of excrement, leftover food, and other debris will help to prevent bacterial and fungal growth. To keep the substrate odor-free and clean, spot clean it frequently and change it completely every few months.

Keeping an eye out for symptoms of illness: Keep a regular eye out for any indications of disease or injury in your tortoise, such as alterations in appetite, behavior, or look. Keep a watchful look out for any anomalies or indications of infection on their limbs, nose, mouth, and eyes.

- Indices of Illness

Early detection and treatment of sickness in African spurred tortoises depend on the ability to identify symptoms. Symptoms might differ based on the underlying medical condition, however some typical indicators of illness to look out for are:

- Diminished hunger or reduction in weight
- Weakness or lethargic mood
- breathing problems or wheezing
- red or swollen eyes
- Nasal secretions or sneezing
- anomalies or discolorations on the shell
- alterations in conduct or degree of activity

It's critical to get your tortoise into the vet as soon as possible if you observe any of these indications or symptoms in them so they can receive a proper evaluation and treatment.

- What to Do If Your Tortoise Needs Medical Care

It's critical to respond quickly to your African spurred tortoise's medical needs if you believe they are ill or injured. If your tortoise needs medical attention, take these actions:

Make contact with a qualified veterinarian who specializes in treating African spurred tortoises. Make an appointment as soon as feasible. Give specific details regarding the symptoms, behavior, and any recent dietary or environmental changes that your tortoise has experienced.

Get ready for the vet visit by gathering any pertinent details for the veterinarian to know, including as your tortoise's nutrition, housing circumstances, and medical history. To avoid harm or escape during transit, place your tortoise in a safe carrier or container.

Observe the veterinarian's advice: To guarantee the best possible outcome for your tortoise, heed the veterinarian's advice regarding diagnostic testing, treatment, and aftercare. As directed by your veterinarian, give your tortoise's medication and keep a close eye out for any negative side effects.

Provide supportive care: with addition to receiving veterinary care, provide your tortoise supportive care to aid with their recovery. Some examples of this care include keeping their surroundings tidy and comfortable, providing a healthy diet, and avoiding handling or stress until they recover completely.

Proactive care, consideration for their special needs, and quick intervention in the event of a health crisis are all necessary to preserve the health and wellbeing of African spurred tortoises. You can maintain the health and happiness of your tortoise for many years to come by being aware of common health issues, taking preventative measures, keeping an eye out for symptoms of sickness, and getting veterinarian care when necessary. To guarantee your tortoise receives the care and attention it needs to flourish, don't forget to offer appropriate housing, nourishment, and hygiene. You should also build a relationship with a skilled veterinarian for reptiles.

Chapter 5

African Spurred Tortoises: Breeding Considerations: Offspring Care and Reproduction

The exciting but intricate process of breeding African spurred tortoises involves meticulous planning, preparation, and devotion to guarantee the health and wellbeing of the adults and their progeny. Breeders need to be aware of the distinctive reproductive patterns and needs of African spurred tortoises, one of the largest kinds of tortoises. To assist breeders in effectively producing and raising healthy offspring, we thoroughly examine the reproductive biology of African spurred tortoises, breeding considerations, nesting and incubation strategies, and hatchling care in this complete handbook.

- Biology of Reproduction

Geographically known as Geochelone sulcata, African spurred tortoises normally achieve sexual maturity between the ages of 15 and 20. However, individual maturity might vary depending on environmental factors, genetics, and food. Females have a flat plastron and a shorter, narrower tail, whilst males have a concave plastron (bottom shell) and a longer, thicker tail.

The rainy season, which varies depending on the locale, is when African spurred tortoises normally breed. Males engage in complex mating displays, such as head bobbing, circling, and vocalizations, to entice females in their natural habitat. After mating, the female will excavate a shallow nest in sandy substrate or dirt, usually in a sunny spot, and lay a clutch of eggs.

Larger females usually produce larger clutches, however the number of eggs in a clutch can vary greatly, ranging from 10 to 30 eggs or more. To keep the eggs safe from predators and to maintain the ideal humidity and temperature for incubation, the female covers her eggs with soil and meticulously packs the nest.

- Breeding-Related Issues

To guarantee the wellbeing of the breeding adults and their progeny, it is crucial to take into account a number of criteria prior to attempting to breed African spurred tortoises. Some things to think about when breeding are as follows:

Finding compatible people to breed: Look for healthy, unrelated breeding partners who complement each other's sizes and temperaments. To avoid transferring unwanted features to their progeny, avoid breeding people with known health problems or genetic abnormalities.

Providing suitable nutrition and husbandry: To maintain reproductive health and fertility, make sure breeding adults are given a wholesome meal, enough illumination, and suitable shelter. Give them access to clean water, a balanced diet of grasses, leafy greens, and occasionally treats, as well as the right amount of humidity and temperature.

Watching for indicators of reproductive behavior: Keep an eye out for indicators of reproductive behavior in adults who are breeding, such as heightened activity, displays of courtship, and nesting habits. During the breeding season, introduce males and females and keep a close eye on their interactions to create opportunities for mating.

Replicating natural nesting conditions: Provide a deep layer of sandy soil or substrate for females to dig their nests in order to provide a suitable nesting space within the breeding enclosure. To promote natural nesting behavior, make sure the nesting location is shielded from distractions and receives enough sunshine.

- Techniques for Nesting and Incubation

To guarantee the health and viability of the eggs, it is crucial to closely monitor the nesting place and provide the right conditions for incubation after a female African spurred tortoise has laid her eggs. To increase the likelihood of a successful hatch, use these nesting and incubation techniques:

Locate the nesting spot and mark it carefully to avoid inadvertently disturbing the eggs or causing damage to them. Mark the border of the nest with flags or stakes, and keep a vigilant eye out for any indications of digging or disturbance.

Providing the right nesting substrate: To make it easy for females to dig their nests and to maintain the right humidity and temperature levels for incubation, make sure the nesting substrate is deep, loose, and well-drained. To produce the perfect nesting habitat, use commercial reptile incubation substrate or a blend of sand and soil.

Temperature and humidity monitoring: Keep an eye on the nesting area's temperature and humidity levels to make sure they stay within the ideal range for egg incubation. Track ambient conditions with a digital thermometer and hygrometer, and make necessary adjustments to ensure stability.

Defending the nest: Cover the nest with hardware cloth or wire mesh that is fastened with rocks or pegs to keep it safe from predators, bad weather, and other disruptions. Make sure the covering keeps out predators and permits sufficient ventilation and sunshine penetration.

Egg incubation: After the eggs are laid, if desired, carefully remove them from the nest and place them in an artificial incubator for regulated incubation. Use a DIY incubation chamber with a consistent humidity and temperature control system, or use a commercial reptile egg incubator.

- Taking Care of Hatchlings

To guarantee the health and survival of the hatchlings, it is crucial to provide them with the appropriate care and husbandry when the eggs hatch. Care for hatchlings of African spurred tortoises by following these guidelines:

Creating an appropriate enclosure: Provide the hatchlings with a small, escape-proof habitat that is the right temperature, humidity, and substrate. For thermoregulation, give hiding places and basking areas in addition to using a shallow dish for drinking water.

Feeding a nutritious diet: To satisfy the hatchlings' nutritional demands, provide a diversified diet that consists of leafy greens, finely chopped grasses, and little amounts of vegetables. Use vitamin D3-containing calcium powder to dust food to promote the development of strong bones.

Keep an eye out for signals of growth and development: Keep a watchful eye out for indicators of the hatchlings' development, such as weight gain, shell growth, and activity level. Make sure they have routine veterinarian examinations to make sure they are healthy and free of parasites or other medical conditions.

Socialization and handling: To assist the hatchlings become used to human contact and lessen stress, handle them gently and frequently. To foster confidence and trust, use mild praise and favorite foods as forms of positive reinforcement.

Introduce the hatchlings to a larger outdoor enclosure with access to natural sunlight, grazing possibilities, and enrichment items gradually when they have grown bigger and stronger. To guarantee their safety and wellbeing, keep a tight eye on their behavior and offer supervision.

Careful planning, preparation, and commitment are necessary while breeding African spurred tortoises in order to protect the health and welfare of the breeding adults as well as their young. Breeders can successfully create and rear healthy young by comprehending the reproductive biology of African spurred tortoises, putting appropriate breeding considerations into practice, and adhering to nesting, incubation, and care practices. To guarantee the greatest results from your breeding efforts, always keep in mind to put the health and wellbeing of the animals first at every stage of the process. You can also seek advice from knowledgeable breeders or doctors who specialize in reptiles as needed.

Chapter 6

Activities for Enrichment: Maintaining a Happy and Stimulated Tortoise

For African spurred tortoises to remain mentally engaged, physically active, and generally content in captivity, enrichment activities are crucial. Even while these reptiles might not be as social as some other pets, they nevertheless gain from having opportunity to interact with their surroundings, engage in natural activities, and encounter various stimuli. In order to improve the general wellbeing and quality of life of African spurred tortoises, we examine a variety of enrichment techniques and activities in this extensive guide, such as habitat improvements, foraging possibilities, social interactions, and sensory stimulation.

- Enhancements to Habitat

Improving the African spurred tortoise's habitat is a good method to give these reptiles enrichment and make their surroundings more interesting. You can promote exploration, movement, and natural behaviors by including naturalistic aspects, hiding places, climbing structures, and other characteristics. Some improvements to the habitat to think about are:

Adding hiding places: Give tortoises a range of places to hide, such half-logs, plant pots, and rock caverns, where they can hide and feel safe. Hiding places provide safety and seclusion, lowering tension and encouraging natural tendencies.

Building climbing structures: To build climbing structures and sunbathing spots inside the enclosure, use branches, rocks, and other natural materials. Opportunities for climbing give tortoises the ability to move, investigate vertical areas, and control their body temperature.

Providing basking platforms: To create warm, dry locations where tortoises can bask and control their body temperature, provide elevated basking platforms or shelves under heat lamps or ceramic heat emitters.

Building tunnels and burrows: To build tunnels and burrows for tortoises to explore and hide in, add PVC pipes, hollow logs, or tunnels constructed of natural materials. The ability to burrow enables tortoises to carry out their innate digging habits and build their own protected areas.

- Possibilities for Foraging

For African spurred tortoises, promoting natural foraging activities is a crucial part of enrichment since it enables them to participate in an essential activity from their repertoire of natural behaviors. Giving tortoises the chance to find and eat food stimulates their senses, encourages movement, and enhances brain stimulation. Here are a few foraging chances to think about:

Disperse food items, including chopped vegetables, leafy greens, and edible weeds, throughout the enclosure to encourage the tortoises to look for and investigate food sources. This technique is known as scatter feeding. Dispersed feeding promotes movement and emulates natural foraging practices.

Using puzzle feeders: Provide food in puzzle feeders or enrichment items made to test tortoises' cognitive abilities and promote problem-solving techniques. PVC pipe, plastic balls with holes, or pre-made enrichment toys can all be used to create puzzle feeders.

Food can be buried in natural substrates to entice tortoises to dig and look for it. Examples of such products include leafy greens, veggies, and treats. Food buried in the ground gives mental stimulation and encourages natural digging tendencies.

To keep tortoises entertained and avoid boredom, alternate the kinds of food and enrichment devices that are provided to them on a regular basis. Their senses are stimulated and their interest in foraging activities is maintained when novel textures, odors, and flavors are introduced.

- Social Exchanges

African spurred tortoises may benefit from restricted social contacts with conspecifics under regulated circumstances, despite not being social creatures in the same sense as mammals or birds. Companion matching or letting tortoises engage in supervised play sessions with one another can offer social enrichment and stimulation. Consider the following options for social interaction:

Assigning compatible individuals to pairs: To promote social interactions and companionship, introduce compatible tortoises to one another under supervision. If you see any indications of hostility or stress in their behavior, keep a close eye on them and separate the parties as needed.

Set up supervised playdates: To offer social stimulation and enrichment, set up supervised playdates with other tortoises or pets, such dogs or cats. Make sure that interactions are closely monitored in order to avoid mishaps or harm.

Provide visual obstacles: To establish distinct regions within the cage where the tortoises can see each other but not directly interact, use visual barriers like plants, pebbles, or dividers. Social stimulation is possible behind visual borders without running the danger of violence or territorial behavior.

Rotate playmates: If you are keeping many tortoises together, make sure to alternate playmates on a regular basis to avoid confrontations and give everyone a chance to socialize. Keep a watchful eye on their actions and step in if you notice any symptoms of stress or hostility.

- Perceptual Arousal

African spurred tortoises benefit greatly from sensory stimulation because it gives them meaningful opportunities to see and engage with their surroundings. Giving them opportunities for tactile, olfactory, visual, and aural stimulation can improve their general health and quality of life. Consider the following exercises for sensory stimulation:

Offering a range of textures: To pique curiosity and engage the senses, offer a range of substrate textures inside the enclosure, such as sand, soil, and leaf litter. Because they can feel diverse textures while they walk, burrow, and dig, tortoises are tactilely stimulated.

Giving the tortoises access to natural sights and sounds: Position the enclosure where the tortoises can enjoy views of the sun, breeze, and chirping birds. Natural stimuli encourage sensory stimulation and can make the environment more captivating.

Adding new scents: To pique the tortoise's sense of smell, provide new scents to the surrounding area, such as freshly cut fruit, flowers, or herbs. To promote olfactory exploration, distribute fragrant flowers or plants throughout the enclosure and/or provide scented snacks.

Providing tactile enrichment resources: Let tortoises explore and engage with natural branches, rocks, and plant materials as examples of tactile enrichment materials. Use a variety of sizes, shapes, and textures to promote tactile engagement and stimulation.

Offering enrichment activities to African spurred tortoises is crucial to fostering both their physical and emotional health while they are kept in captivity. Owners of tortoises may provide a more stimulating and engaging environment for their pets by including social interactions, foraging opportunities, habitat modifications, and sensory stimulation into their daily routine. Always pay close attention to your tortoise's behavior and adjust enrichment activities to suit their unique requirements and preferences. African spurred tortoises can live long, robust lives and prosper in captivity for many years if given the right enrichment.

Chapter 7

Responsible Ownership and Conservation Efforts: Legal and Ethical Considerations

It is crucial for owners and lovers of African spurred tortoises to comprehend and abide by the ethical and legal guidelines guiding their conservation and upkeep as stewards of the species. Although many people around the world keep these reptiles as pets, habitat loss, illegal trading, and other human activities pose dangers to their natural habitat. This extensive handbook covers appropriate pet ownership, the legal and ethical foundation for African spurred tortoise ownership and conservation, and initiatives to prevent the exploitation and decrease of wild populations.

- Legal Structure

The laws pertaining to the possession and exchange of African spurred tortoises differ from one nation and area to the next. National and international legislation and regulations that safeguard these tortoises are in place in many nations with the intention of preserving their wild populations and controlling their trafficking. The following are some important legal factors to be aware of:

CITES protection: Appendix II of the Convention on International Trade in Endangered Species of Wild Fauna and Flora (CITES) lists African spurred tortoises and governs their international trade, ensuring that it does not pose a threat to wild populations. Permits are needed for the import and export of tortoises and their byproducts, including specimens and shells, due to this listing.

National laws and regulations: To safeguard African spurred tortoises and control their ownership, commerce, and commercial breeding, numerous nations have put national laws and regulations in place. The general goal of these laws, albeit they may differ in their application and enforcement, is to stop the unlawful taking, trading, and use of wild tortoises.

Permits and licensing: In certain countries, obtaining permits or licenses from the appropriate government agencies may be necessary for the ownership, propagation, or trade of African spurred tortoises. These permits may be subject to restrictions and guidelines and may be granted for certain uses, such as scientific research, conservation breeding, or educational initiatives.

Protecting endangered species: Due to habitat destruction, overexploitation, and other threats, African spurred tortoises are deemed vulnerable or endangered in their natural habitat. Because of this, they might be given more protection by national and international regulations meant to preserve endangered species and their natural environments.

- Moral Aspects to Take into Account

When caring for African spurred tortoises, pet owners and enthusiasts should take ethical issues into mind in addition to legal ones. The purpose of ethical considerations is to advance the welfare of individual tortoises and aid in the preservation of their species. These concepts include those of animal welfare, responsible ownership, and conservation ethics. Among the moral things to think about are:

Responsible pet ownership: Taking care of a tortoise demands a long-term commitment and the availability of resources to address its requirements on all levels, including behavioral, emotional, and physical. Owners of pets should make sure they have the skills, time, and finances required to give their tortoise the care and wellbeing it needs for the duration of its life.

Avoiding wild-caught specimens: If at all feasible, pet owners should choose to buy captive-bred African spurred tortoises from respectable breeders or rescue groups rather than wild-caught animals. Because of the stress, trauma, and health problems that come with being captured and transported from the wild, natural populations may be put at further risk by removing captured tortoises from the wild.

Encouraging conservation awareness: People who own pets and those who are passionate about conservation can help spread the word about the dire situation facing African spurred tortoises and other endangered animals.

People can support efforts to safeguard wild populations and their ecosystems by becoming more aware of conservation issues, habitat damage, and illegal commerce.

Encouraging conservation measures: African spurred tortoises and their native habitat can be preserved by funding research programs, conservation organizations, and habitat restoration initiatives. People can support conservation efforts and have a beneficial impact on the survival of threatened species by contributing money, giving their time, or taking part in conservation initiatives.

- Preservation Activities

A variety of tactics are used in the conservation of African spurred tortoises and their environments with the goal of mitigating threats to wild populations and ensuring their long-term survival. Public education, community involvement, captive breeding programs, anti-poaching measures, and habitat protection are a few examples of conservation activities. Among the most important conservation projects and endeavors are:

Protection of habitat: Because habitat deterioration and loss pose serious risks to wild populations, it is imperative that the natural habitats of African spurred tortoises be preserved. To preserve vital habitat for tortoises and other endangered species, conservation organizations strive to create reserves, protected areas, and wildlife corridors.

Anti-poaching measures: Because African spurred tortoises are frequently sought after for their meat, shells, and as pets, they are seriously threatened by both illegal trade and poaching. Conservation organizations fight against the illicit poaching and trafficking of tortoises and enforce laws and regulations meant to protect them in partnership with law enforcement agencies, government authorities, and local communities.

Captive breeding initiatives: These initiatives are essential to the conservation of African spurred tortoises because they lower the market for specimens taken from the wild and increase the number of tortoises kept in captivity in preparation for possible release back into the wild. The major goals of breeding programs are to minimize inbreeding, preserve genetic diversity, and produce healthy progeny for further conservation initiatives.

Public education and outreach: These initiatives encourage good pet ownership behaviors and increase public knowledge of the value of protecting African spurred tortoises and their ecosystems. Aiming to raise public understanding and engagement in conservation, educational initiatives might take the form of media campaigns, school programs, community seminars, and public events.

The long-term existence of African spurred tortoises and other endangered species depends on ethical ownership and conservation initiatives. Pet owners and enthusiasts can improve the wellbeing of individual tortoises and support larger conservation efforts to safeguard wild populations and their habitats by being aware of and abiding by legal and ethical considerations. People may positively impact African spurred tortoise conservation and help preserve the species' survival for future generations by encouraging ethical pet ownership, supporting conservation programs, and raising awareness about conservation concerns.

Chapter 7

FAQs (Frequently Asked Questions) on Petting African Spurred Tortoises

Sulcata tortoises, another name for African spurred tortoises, are interesting reptiles that have gained popularity as pets recently. But taking good care of these tortoises demands understanding, devotion, and understanding of their special requirements. We answer some of the most common queries regarding keeping African spurred tortoises as pets in this extensive guide, which covers subjects including maintenance needs, food, habitat, health issues, and more.

- How long does an African spurred tortoise live to?

In captivity, African spurred tortoises can live for up to 150 years, making them long-lived reptiles. These tortoises require a long-term commitment from pet owners because they can survive for several decades with the right care and maintenance.

- How big can an African spurred tortoise get?

One of the largest types of tortoises are African spurred tortoises, which can grow to be up to 30 inches (76 cm) long and weigh more than 100 pounds (45 kg). Individual tortoises' sizes, however, could differ depending on a variety of factors, including nutrition, genetics, and environmental influences.

- African spurred tortoises consume what?

The primary diet of African spurred tortoises is plant stuff, as they are herbivores. A range of grasses, leafy greens, vegetables, and occasionally fruits should make up their diet. A well-rounded food that satisfies their nutritional requirements must be given, along with calcium supplements to maintain the health of their shells and bones.

- How frequently should my African spurred tortoise be fed?

Fresh greens and vegetables should be the staple food of adult African spurred tortoises, with the occasional fruit added as a treat. Depending on their activity level and appetite, adults may need to eat several times a week, while juveniles may need to be fed more frequently.

- Which kind of enclosure works best for spurred African tortoises?

The large enclosures needed for African spurred tortoises to wander and explore are necessary. With their abundance of natural light, grazing room, and exercise area, outdoor enclosures are perfect. Enclosures inside should be big enough to fit the tortoise comfortably, and they should have the right substrate, lighting, and heating.

- Do UVB lights be necessary for African spurred tortoises?

Certainly, in order to maintain their health and wellbeing, African spurred tortoises need access to UVB lighting. The creation of vitamin D3, which is required for calcium metabolism and shell formation, depends on UVB light. The enclosure's UVB light source aids in the prevention of metabolic bone disease and other health problems.

- How do I keep the enclosure housing my tortoise warm?

In order to thermoregulate, African spurred tortoises need a warm, dry habitat with access to basking spots. 85 to 95°F (29 to 35°C) is the ideal range for daytime temperatures, with a basking spot that can get as high as 100°F (38°C). It can be as low as 70 to 75°F (21 to 24°C) at night.

- Are Spurred African Tortoises Able to Hibernate?

During hot, dry seasons, African spurred tortoises may go through periods of slumber or aestivation in their natural habitat. For tortoises kept in captivity, hibernation is not advised unless the animals are in excellent health and can have their surroundings regularly monitored. Before attempting to hibernate a tortoise, seek advice from a veterinarian with experience caring for reptiles.

- How should I care for my spurred African tortoise?

To reduce stress, it's crucial to handle African spurred tortoises in a calm and kind manner. Avert lifting them by their limbs, tail, or head, and instead support their shell with both hands—one under the plastron, the other holding up the carapace. Make sure that handling sessions are brief and that the turtle has plenty of downtime in between contact.

- Which health conditions are prevalent in African spurred tortoises?

African spurred tortoises frequently suffer from respiratory infections, metabolic bone disease, eye infections, shell damage, and parasitic infections. It's critical to keep an eye out for any symptoms of sickness in your tortoise, including as alterations in behavior, food, or appearance, and to seek veterinary attention if necessary.

For devoted reptile lovers, African spurred tortoises are intriguing and gratifying pets. Pet owners may guarantee the health and well-being of these amazing reptiles for many years to come by being aware of their specific needs, giving them the care and husbandry they require, and responding to often asked questions and concerns. For advice and assistance with any additional queries or worries regarding taking care of African spurred tortoises, speak with a licensed veterinarian or knowledgeable reptile keeper.

Summary

To sum up, African spurred tortoises are amazing reptiles that captivate devoted lovers when kept as pets. We've covered a wide range of topics in this book to assist pet owners in understanding and providing for the needs of their tortoises, including care requirements, diet, housing, health issues, and more.

Pet owners may guarantee the health and well-being of their African spurred tortoises by giving them the right husbandry, which includes a wholesome feed, a large cage with UVB illumination and the right temperature, and routine veterinary check-ups. Furthermore, encouraging responsible pet ownership and supporting initiatives to preserve wild populations depend on people having a thorough grasp of the ethical and legal issues related to their ownership and conservation.

In the end, taking good care of African spurred tortoises takes a sustained investment of time, effort, and understanding of their particular requirements. These amazing reptiles can have long, healthy lives and continue to amaze and delight their owners for many years if given the right care and attention. Consult knowledgeable reptile veterinarians or seasoned reptile keepers if you have any more queries or worries regarding caring for African spurred tortoises. By working together, we can protect these incredible animals' health and welfare and help to preserve them for enjoyment by future generations.